The Super Tiny World

Alec
the Actinomyces

Written and illustrated by Dr. Rochman

To my children,
May your thirst for knowledge
never be quenched.

ISBN: 978-1-7753850-0-4 (e-book)
ISBN: 978-1-7753850-1-1 (hardcover)
ISBN: 978-1-7753850-4-2 (paperback)

AF268904

Microbes

Microbes (or microorganisms*), are very tiny living creatures invisible to the naked eye, such as bacteria, viruses, or fungi. They exist as one cell or in a group of cells. They live in air, earth, water, and even inside plants and animals. The human body is also home to trillions of microbes. Some of them can make you sick, but others are important for your health.

*micro = extremely small, organism = living thing

Bacteria

Virus

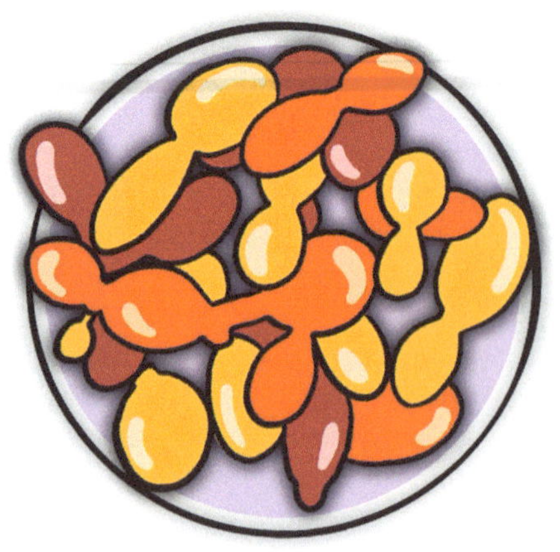

Fungi

Science tools

Scientists use different tools to study microbes.
Two important tools used are petri dish and microscope.

A petri dish is a plastic/ glass dish with lid used by biologists to grow microbes
(white dots = microbes)

A microscope is an object used to see things that are too small for our eyes to see, such as microbes

Hello

I am **Alec the Actinomyces** (ak·tǐ·nō' mī, sēz).
I am a **microbe** who lives **in and around you.**
In this book, I will tell you my story.

This is how I look in **real life** under a microscope. I have **branches**, just like trees.

The family

My family is called the Actinomyces. In Greek, Actis means ray and mykes means fungus (plural: fungi). So our family name means 'ray fungus', due to our structural similarities to Fungi.

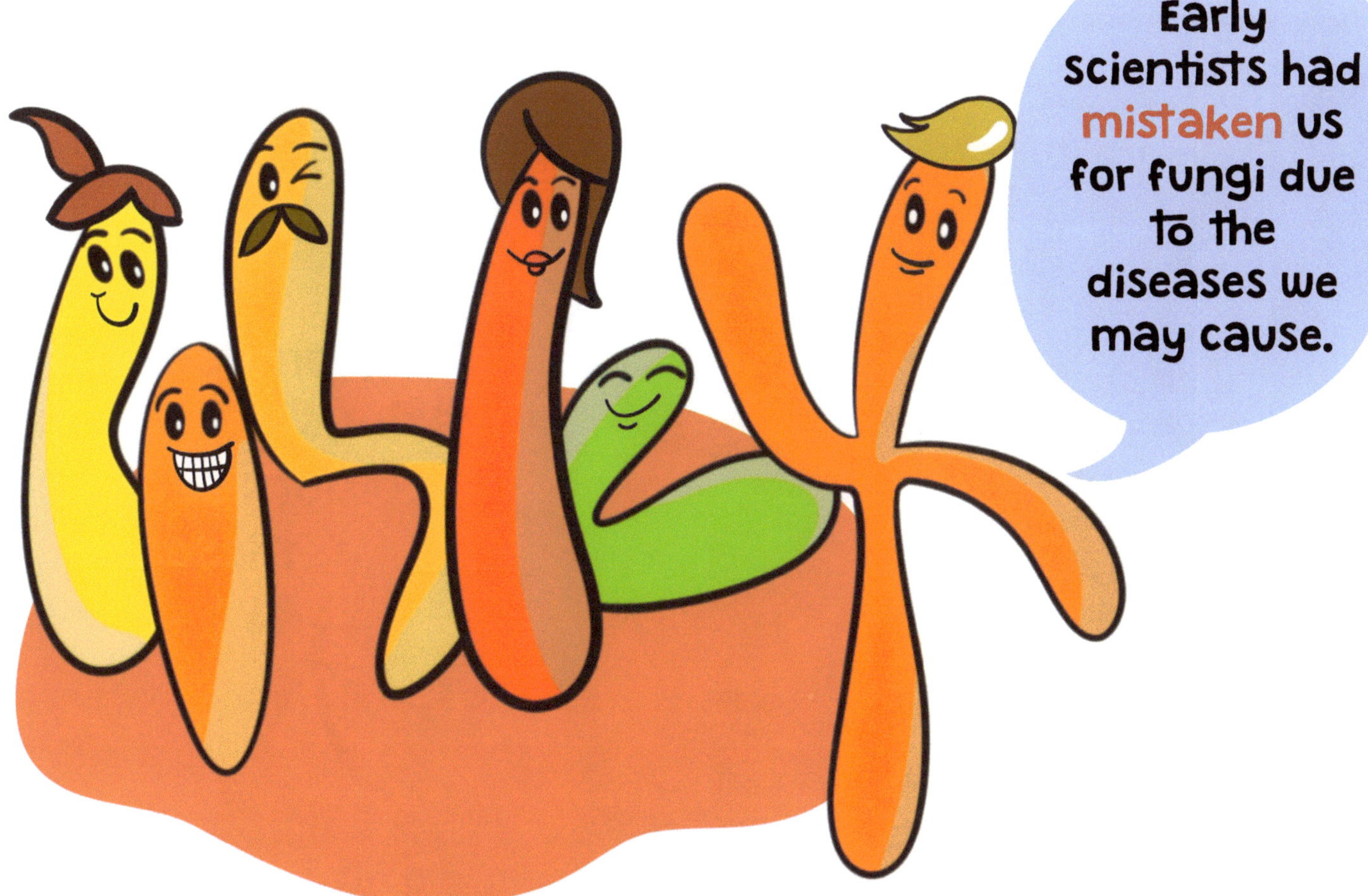

Family colony

Just like you, Actinomyces love to stick together with their families in groups called colonies. Here are the colonies grown on a growth medium in petri dishes.

The picture shows an Actinomyces colony. Spot their **filaments** both in the air and under the surface. From the structure, they look like fungi.

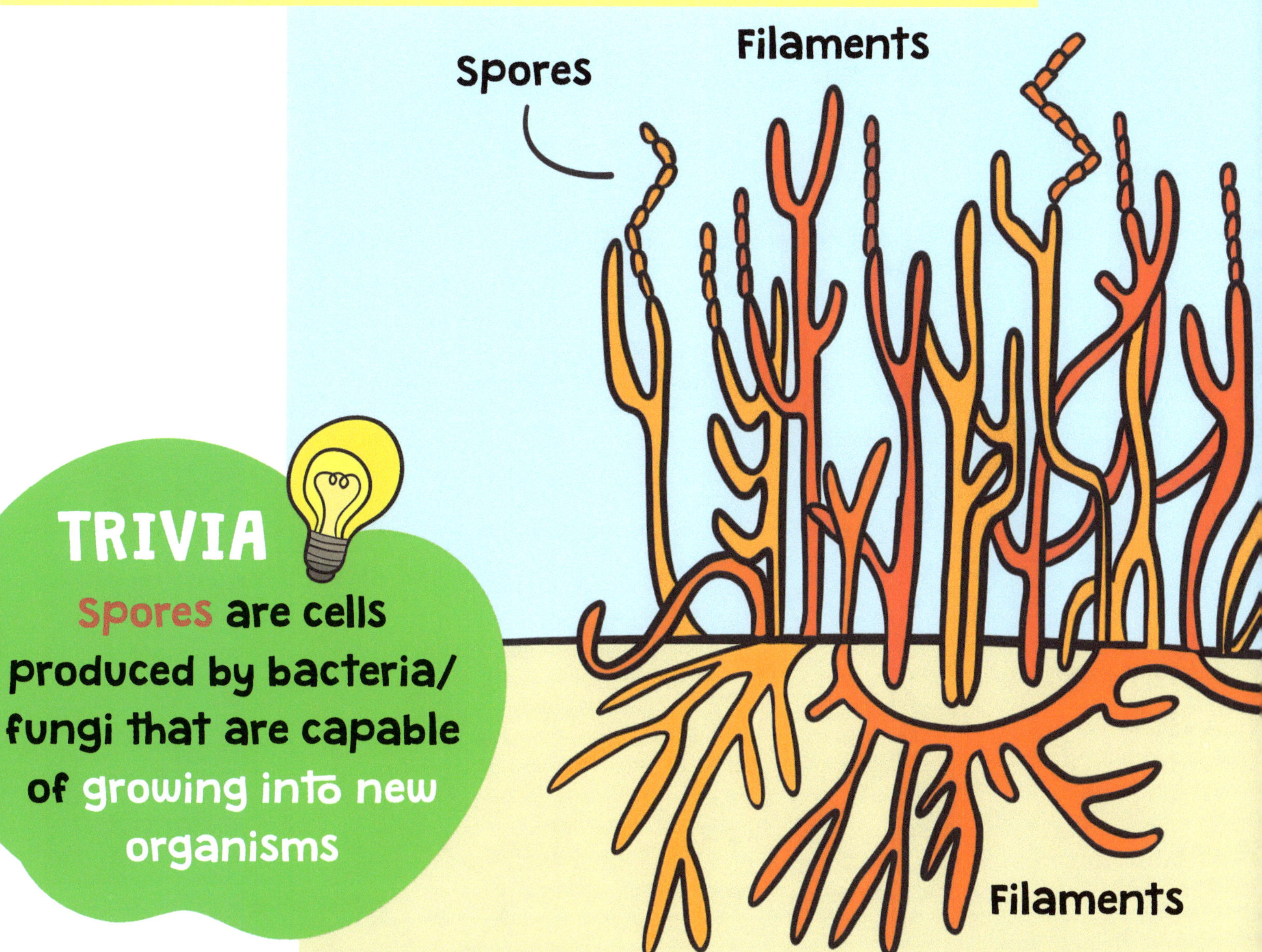

TRIVIA

Spores are cells produced by bacteria/ fungi that are capable of growing into new organisms

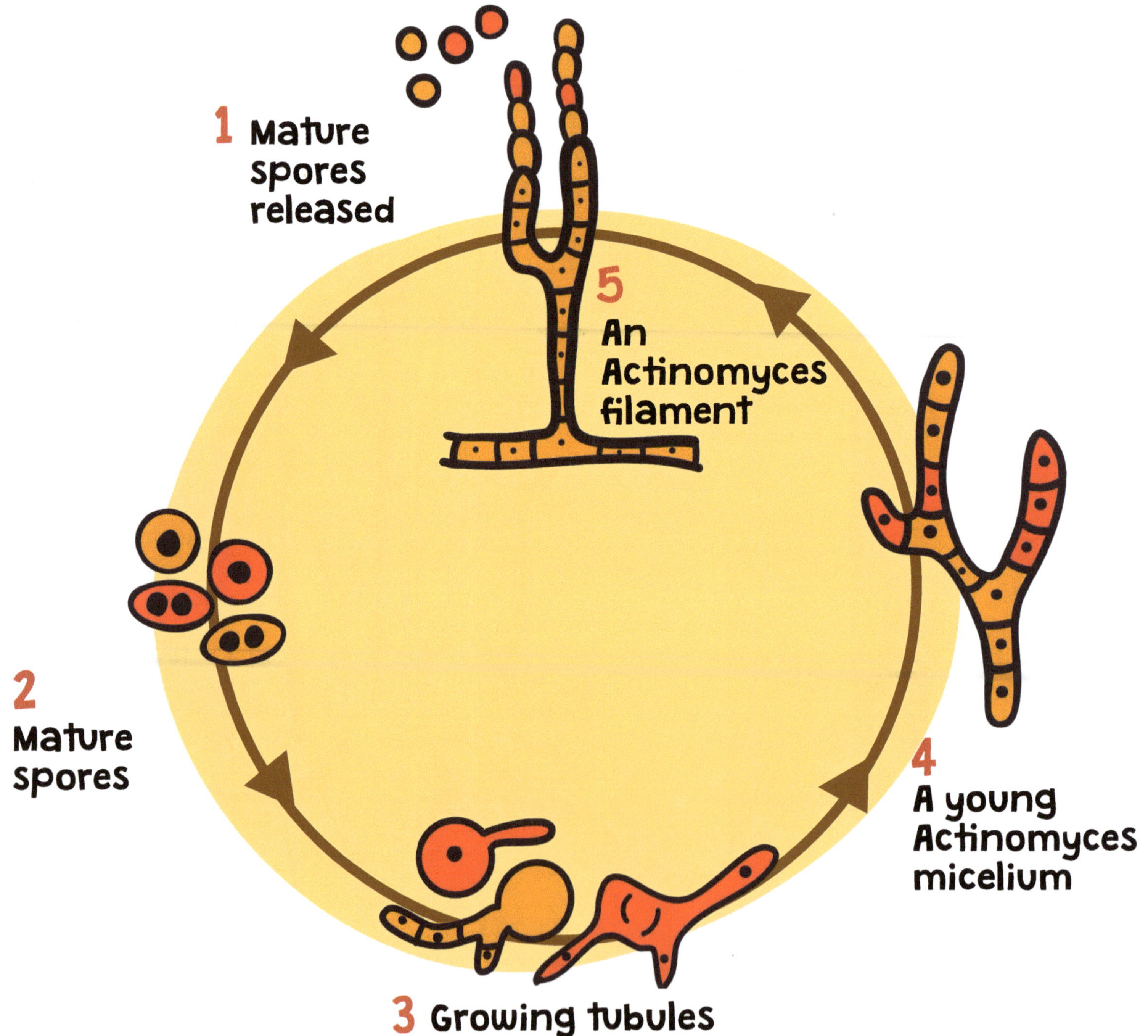

The Life Cycle
1 Mature spores released
2 Mature spores
3 Growing tubules
4 A young Actinomyces micelium
5 An Actinomyces filament

How we breathe

Actinomyces can either breathe with oxygen from the air (aerobic), or without oxygen (anaerobic). Aerobic microbes use oxygen to produce energy, and anaerobic microbes do not need oxygen for energy production.

Where we are

We are found in places you cannot imagine! Our lives are so much fun that you wouldn't want to miss learning about us!

We enjoy living in places that are warm at 20 to 40°C (70 to 100°F), just like at the tropics. If it is too cold or too hot, we won't survive!

One of our favorite places to live is on your teeth. It is warm in your mouth and there are nutrients from food remains that are left uncleaned in between your teeth.

Ouch!
But when you don't clean your teeth well, we may cause you pain.

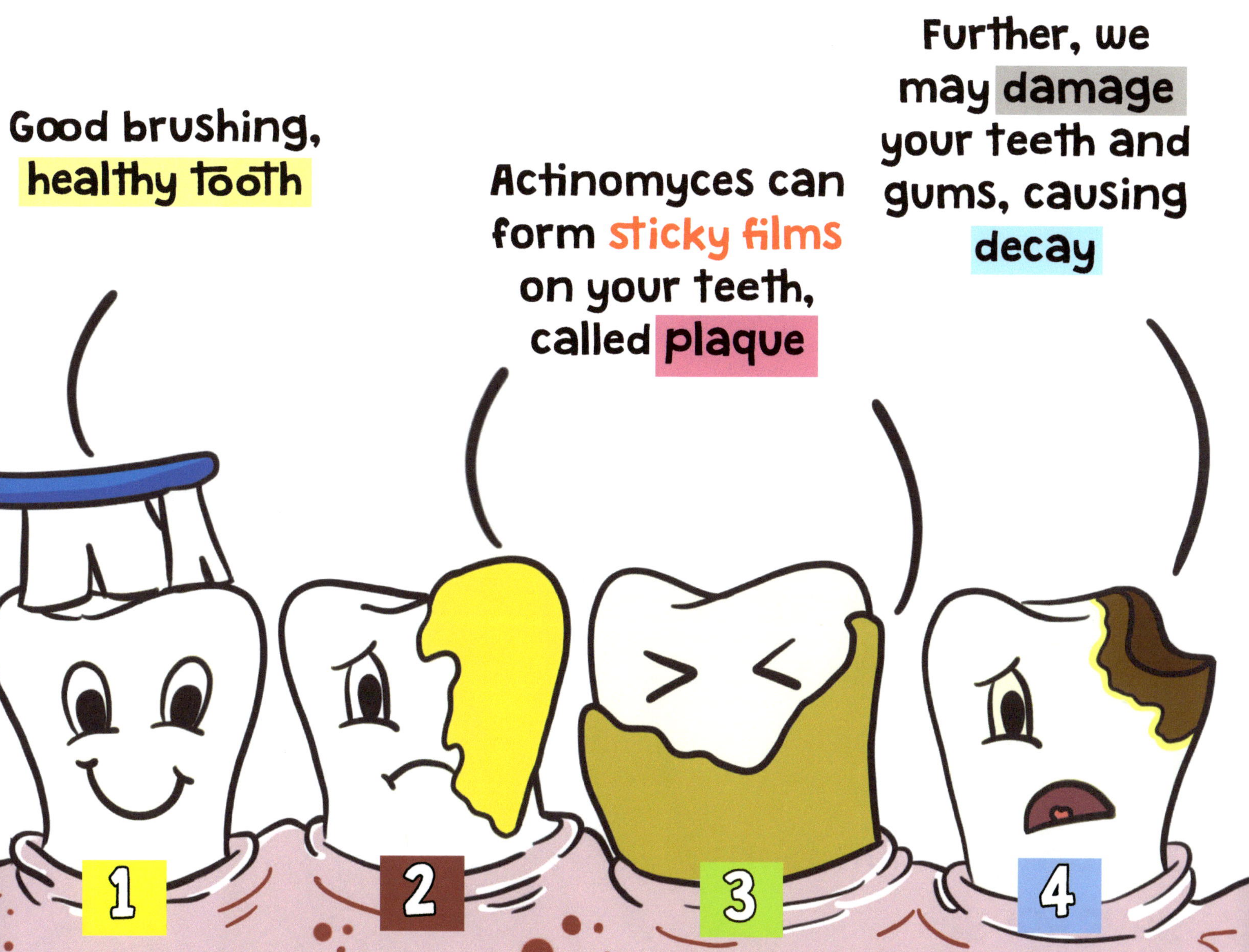

This is what happens in your mouth when you 1) brush your teeth well, 2 and 3) left some food in between your teeth where Actinomyces form plaque, and 4) don't clean your teeth well. That is why it is important to visit the dentist once in every 6 months to avoid plaque and decay.

In the human body systems

Actinomyces can be found in human systems, including the mouth (1), throat (2), intestinal tracts (3), and bladder (4). They do not cause infection, except when there are abnormal wounds in your body.

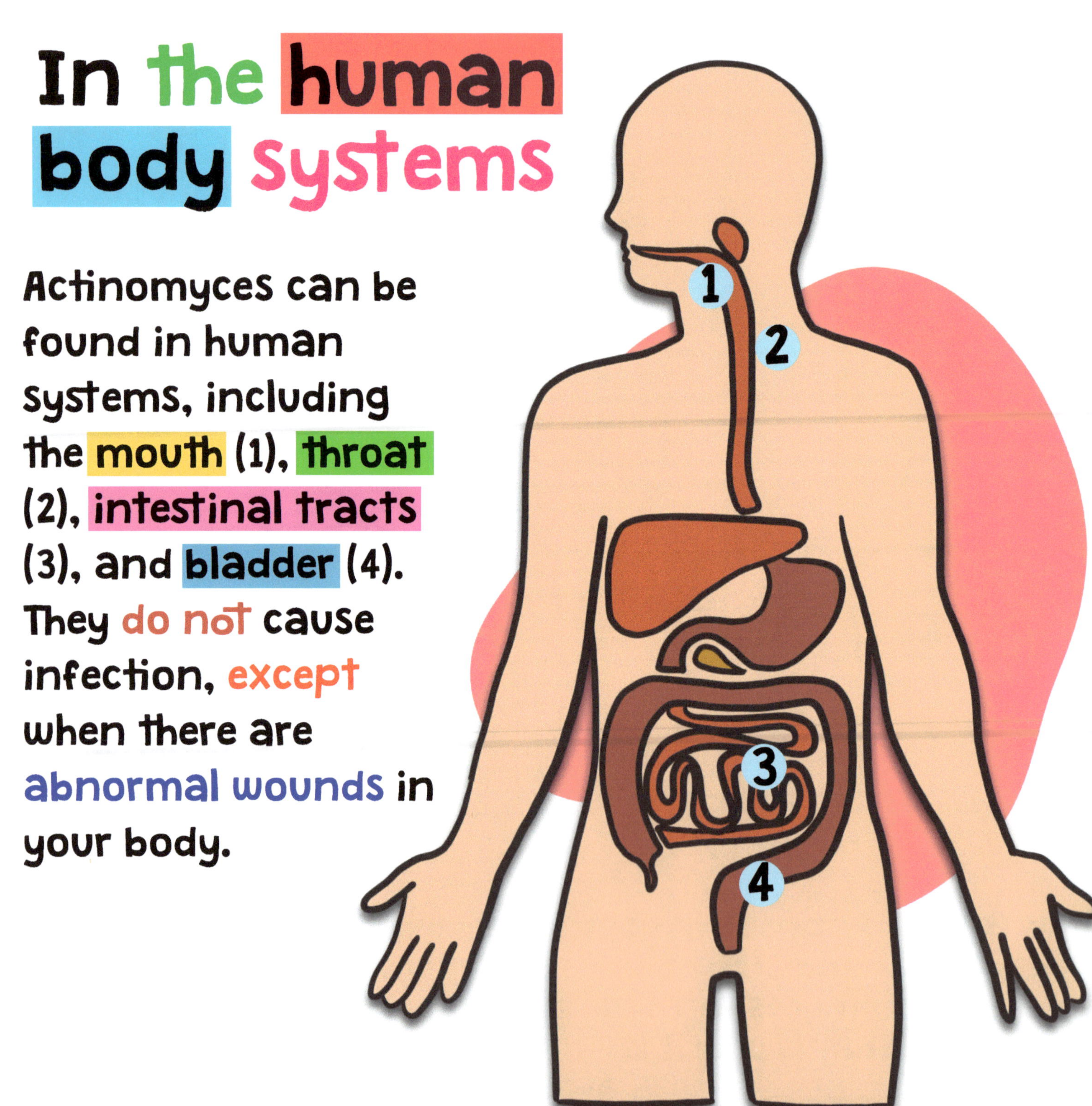

Inside animals

Not only inside humans, we are also common in horses, dogs, and wild animals. We colonize the teeth, mouth, throat, and digestive and genital tracts. We do not normally cause sickness, but we may when there are open wounds.

Microbial infection

Infection happens when microbes make you sick.

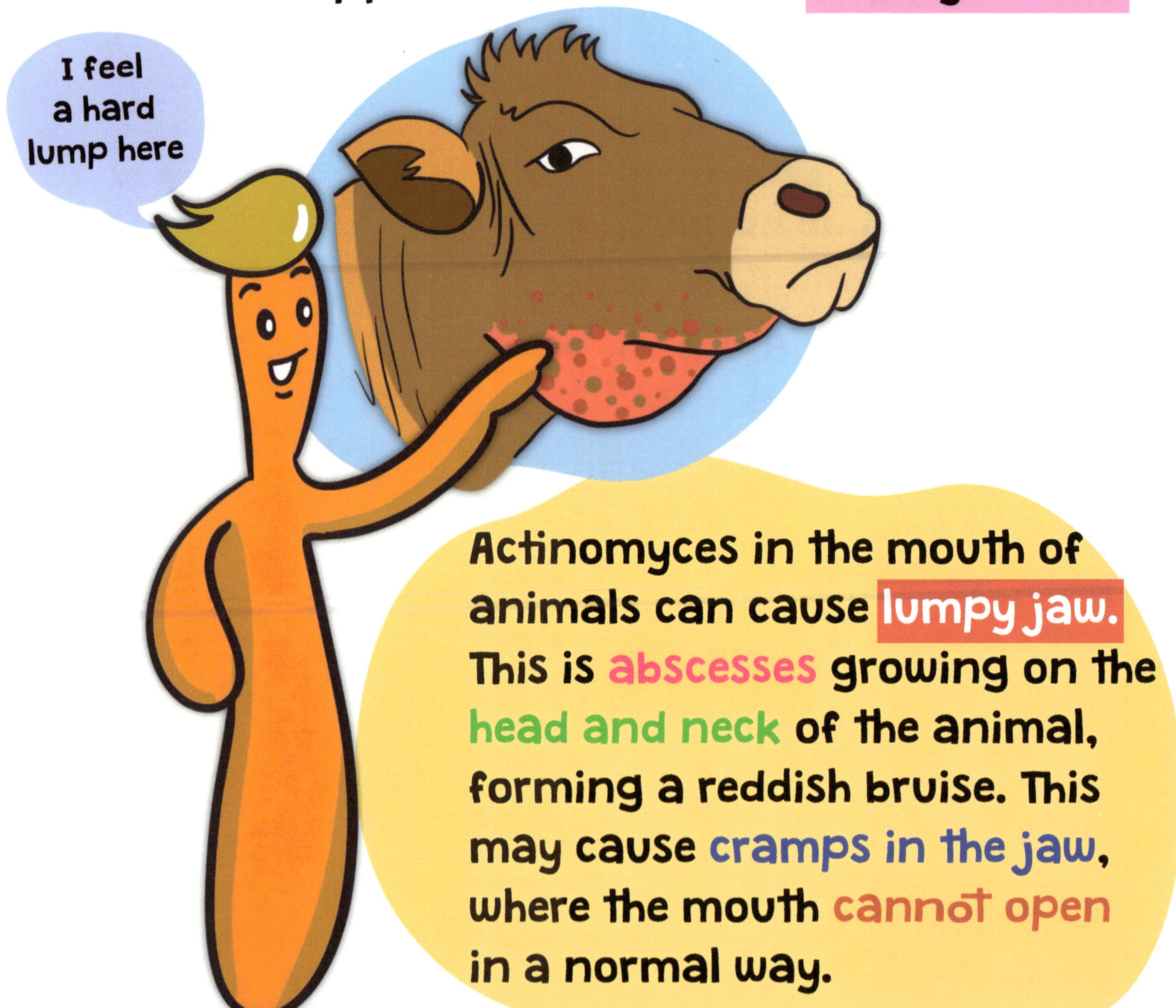

Lumpy jaw treatment

Antibiotics pills, such as penicillin or amoxicillin can drive away Actinomyces. This medication is used to treat lumpy jaw.

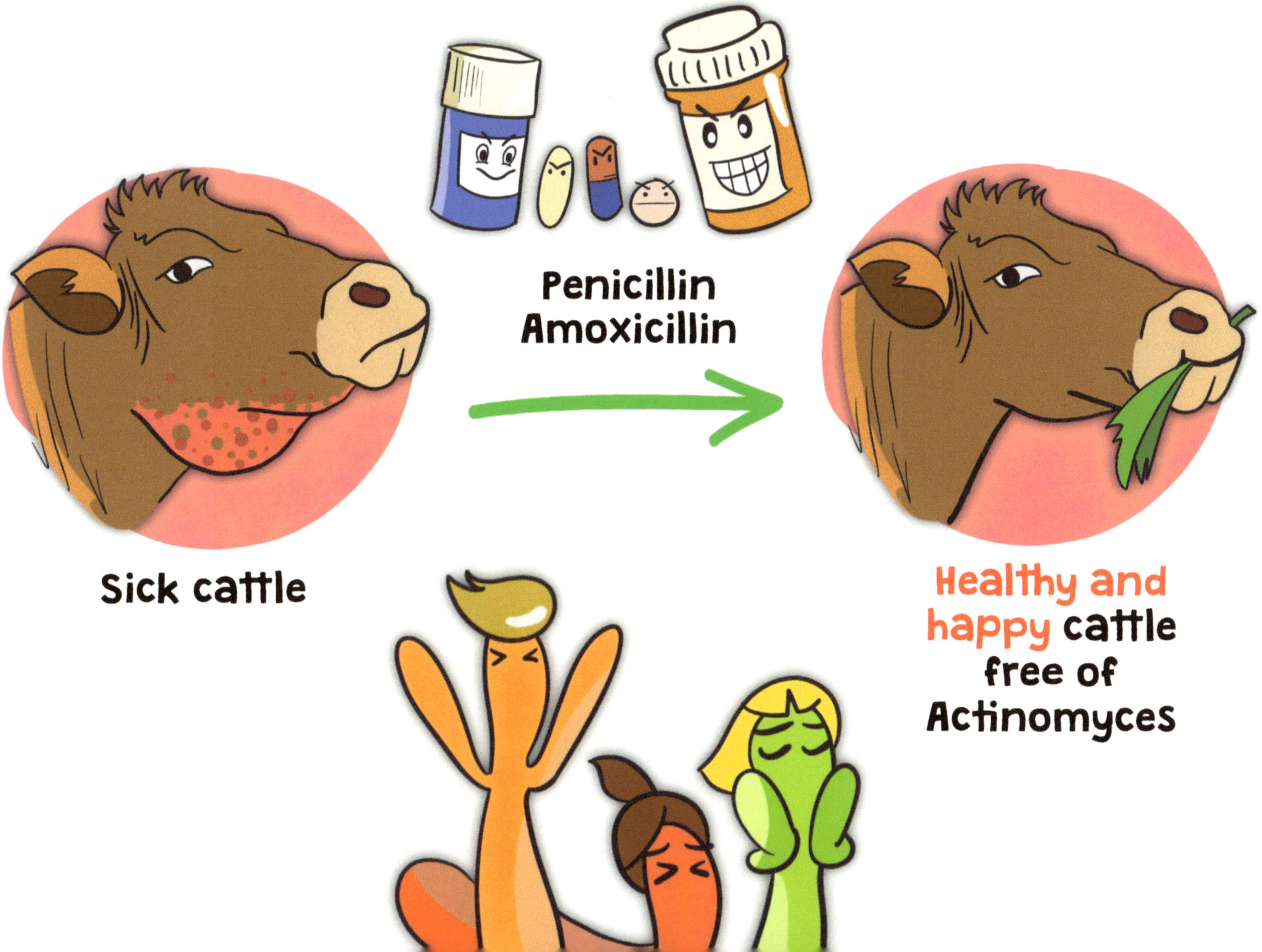

Deep in Soil

Our presence improves soil health and promotes plant growth. We break down complex plant and animal residues and in return give nutrients back to the soil.

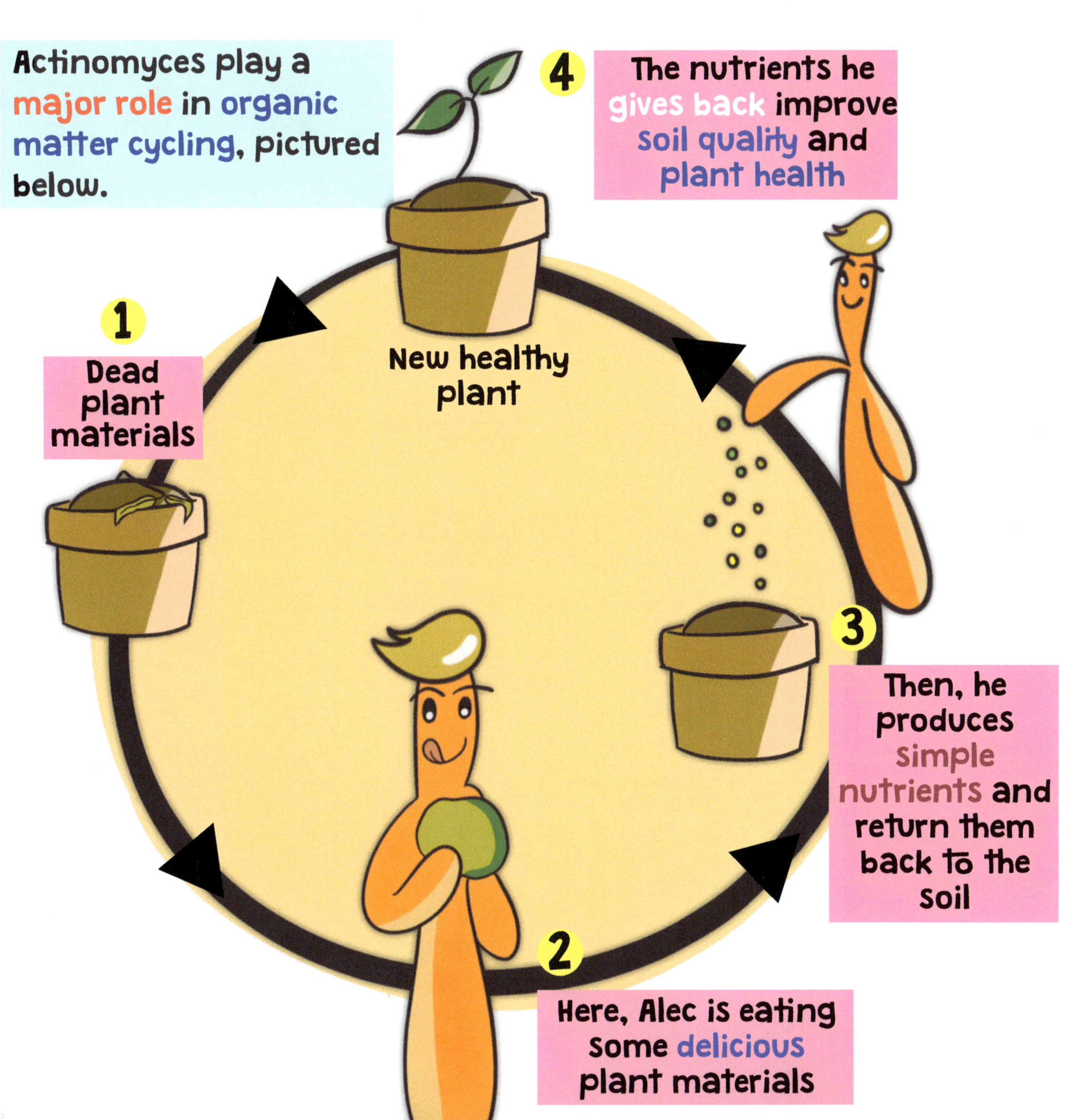

Actinomyces play a major role in organic matter cycling, pictured below.
1
Dead plant materials
New healthy plant
4
The nutrients he gives back improve soil quality and plant health
3
Then, he produces simple nutrients and return them back to the soil
2
Here, Alec is eating some delicious plant materials

Composting

Decomposers are organisms that eat and break down food and plant materials.

Large decomposers: snails, sow bugs, slugs, beetles, ants, flies, and worms. They grind, bite, tear, and chew food scraps into smaller pieces.

Small decomposers: Fungi and Actino- myces. They break down smaller food scraps into simpler nutrients beneficial for soil. This process is called composting.

How **composting** is *done* by Actinomyces

This is Alec's sister. She loves munching tough pieces of food scraps. This helps composting.

Compost is decayed organic materials (plant, animal, food scraps) used as plant fertilizers.

Compost rich in nutrients promotes plant growth. Actinomyces in soil means composting success!

The Smell of earth

Do you know what is behind the fresh and earthy smell after first rain? It is a substance called geosmin, that we produce with other soil microbes.

Pesticides

Pesticides are substances that are used to control pests on plants. They help protect plants from unwanted weeds, fungi, or insects.

Actinomyces produce special substances that repel mosquitoes and rodents.

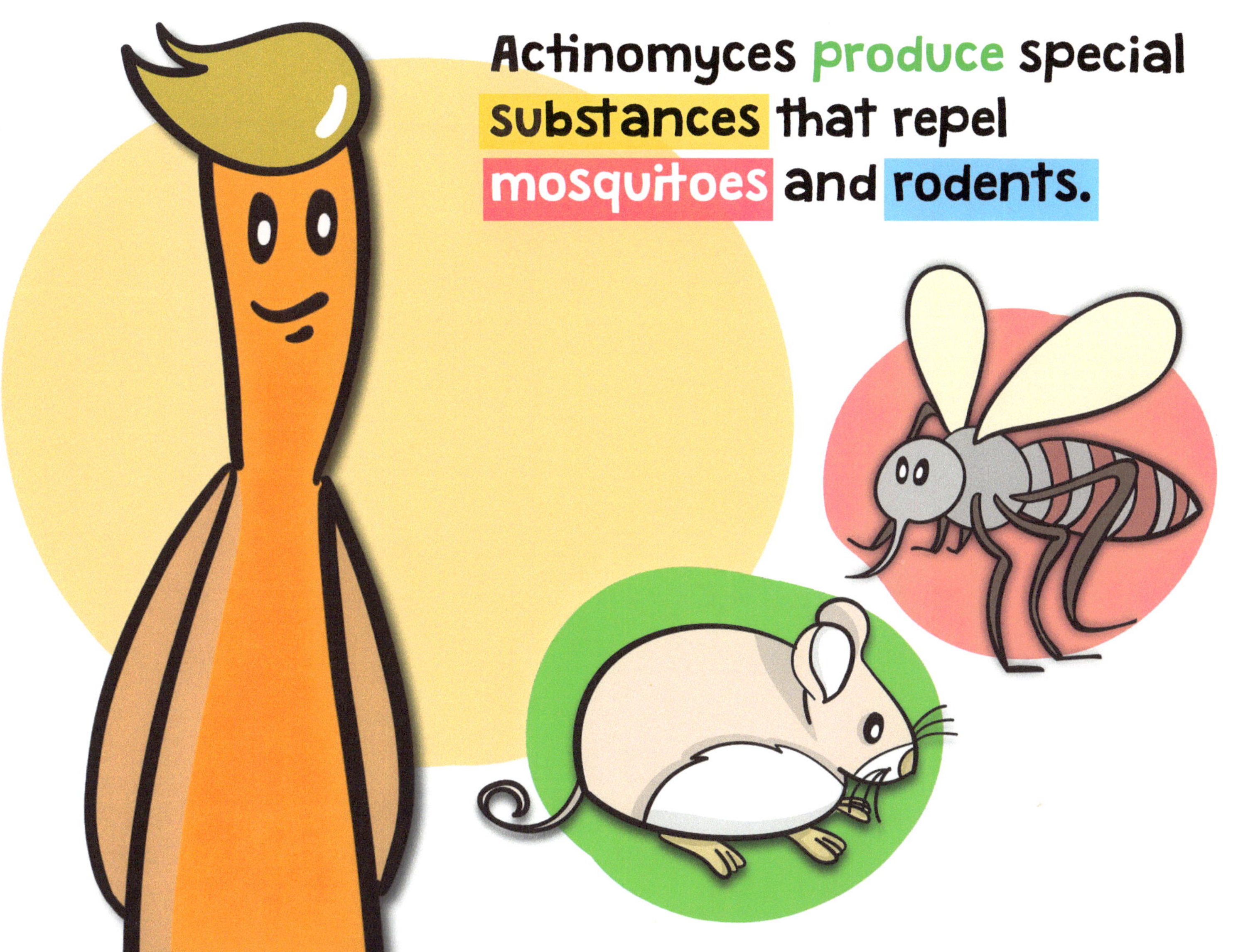

POP Quiz

See how well you understand the Actinomyces!

1. Name 3 types of microbes.
2. What should you do to avoid having Actinomyces on your teeth?
3. Where might you find Actinomyces in your body?
4. When Actinomyces invade cattle, what kind of sickness might it cause?
5. What medication can drive Actinomyces away from sick animals?
6. What can help protect plants from unwanted weeds, fungi, or insects?

Answers: 1) Bacteria, virus and fungi. 2) Brush your teeth. 3) Mouth, throat, intestinal tracts, and bladder. 4) Lumpy jaw. 5) Antibiotics. 6) Pesticides.

Good job!

Brace yourself, the busy life of Billy the Bacillus is coming next! Just as fun as Alec's with some more exciting facts to learn!

References:
1) WHO Model Formulary 2008 (PDF). World Health Organization. 2009.
2) Pubmed Health. What are microbes? www.ncbi.nlm.nih.gov.